Energy
234

鸟中宇航员
The Astronaut Bird

Gunter Pauli

[比]冈特·鲍利 著

[哥伦]凯瑟琳娜·巴赫 绘

贾龙慧子 译

上海远東出版社

特别感谢以下热心人士对童书工作的支持：

匡志强　方　芳　宋小华　解　东　厉　云　李　婧
刘　丹　熊彩虹　罗淑怡　旷　婉　杨　荣　刘学振
何圣霖　王必斗　潘林平　熊志强　廖清州　谭燕宁
王　征　白　纯　张林霞　寿颖慧　罗　佳　傅　俊
胡海朋　白永喆　韦小宏　李　杰　欧　亮

目录

Contents

一只鹤正准备飞越喜马拉雅山。他看到一只雁也正准备翻山呢。

鹤便提议道：“嗨！咱们一起飞吧，这样也容易避开捕食者。”

雁回应道：“谢谢你的提议，这主意不错，有个同伴陪着总是好的。你飞得很高，对吧？”

A crane is getting ready to fly across the Himalayas. He spots a goose preparing to travel the same route.

"Hey, let's fly together and be safe from predators," the crane suggests.

"Thanks. That does make sense, and some company is always welcome. You fly very high, don't you?"

他看到一只雁……

He spots a goose ...

飞得越高，空气越稀薄

The higher we go, the thinner the air

鹤说：“是的，飞得高更安全些。老鹰可不能像我一样在一万米的高空飞翔。我可以安全地飞到我的繁殖地去，不过得先爬升到那个高度才行。”

雁说道：“我可不想飞那么高。飞得越高，空气越稀薄，就得消耗更多的力气。但为了到达高山另一边那些肥沃的土地，我也准备好尽最大的努力了。”

“Yes, I do so to be safe. No eagle can fly at ten thousand metres like I do. Up there I am safe when travelling to my breeding grounds, but I have to get to that altitude first.”

“I don’t want to go quite that high. The higher we go, the thinner the air, which means I have to work harder. But I am prepared to go to great effort to get to the land of plenty on the other side of these huge mountains.”

鹤说道：“为了找到能安全产卵孵蛋的地方，我们得搞清楚怎么翻过这些大山。”

雁说：“问题在于怎么为我们的肌肉和大脑供给充足的氧气。”

鹤说道：“正是如此。海拔高的地方供给身体的氧气的确少一些，但我们能搞定这个问题。”

“For us to get to where we can lay our eggs and have our chicks hatch safely, we need to know how to cross these mountains.”

“The challenge is how to get enough oxygen supply to our muscles and brain.”

“As it is, there is less oxygen up there to power the body, but we manage well.”

安全孵蛋

Ours chicks hatch safely

……人们行走的长队。

... the long queues of people walking.

雁问道：“你知道吗？人类没法达到我们飞行的高度，绝大多数的其他鸟类也做不到。”

“不过人类都是用走的，不是飞的……”鹤说。

“他们的确靠走路。你没见过人们在世界最高的山上行走并且试图登顶的长队吗？”

“Did you know that the two of us are able to do what no human can? And what very few other birds can do?”

“But people are meant to walk, not fly…” Crane says.

“And walk they do. Have you not seen the long queues of people walking on the highest mountain in the world, trying to reach the top?”

“我见过，但是人类的行为很奇怪。他们只在乎爬上山峰，一点也不在乎他们一路上遗留的垃圾。而且他们的乐趣在于拍照，以此证明自己攀上了高峰。”鹤说道。

“人类需要几天甚至几周的时间来习惯高处稀薄的空气，而且需要在这种空气下行走相当长的距离。咱们几个小时就能搞定。”雁说道。

“I have. People do act strangely, though. Walking up here, not caring how much trash they leave behind. And very interested in taking pictures of themselves reaching great heights!”

“It takes people days, or even weeks, to get used to the thin air up here, and cover great distances. We do it in a matter of hours,” Goose says.

……乐趣在于拍照……

… interested in taking pictures …

……秃鹫可以飞得更高。

... vultures can fly higher.

“咱们挺厉害的，对吧？我觉得也就秃鹫可能比我们更胜一筹。”

“你说得对。在稀薄的空气中，秃鹫可以飞得比任何其他鸟类更高更远，但我们还是可以自夸一下的。”

“妈妈教过我呼吸的时候要快速吸气，缓慢吐气，然后保持这个节奏。”

“呼吸是最重要的。我们的心脏拥有上百万比头发丝还细的小动脉，来向身体运输富含氧气的血液，在我们需要的时候为我们提供能量。”

“Aren’t we good? I believe only the vulture does better.”

“You are right, vultures can fly higher, and for longer in thin air, than any other bird. But we can still be proud of ourselves.”

“My mother taught me to breathe in quickly and breathe out slowly, and to keep up that rhythm.”

“It is all in the breathing. Our hearts have millions of small arteries, thinner than a hair, to quickly send oxygen-rich blood around, providing us with all the energy we need – where we need it.”

“大雁啊，告诉我你吃什么能长得这么壮？”

“我有草吃就很开心了。”

“真的？我得吃更多东西来囤积能量，所以起飞之前会吃老鼠、螃蟹甚至小型鸟类。”鹤说道。

“Tell me, Goose, what do you eat that makes you so strong?”

“Oh, I am happy feeding on grass.”

“Really? I need more, so will eat mice, or crabs, even small birds before I set off.”

我有草吃就很开心了。

I am happy feeding on grass.

我会在破晓时分出发。

I take off at dawn.

“我不知道你喜欢什么时候出发，不过我会在破晓时分出发。然后我会径直从海洋飞到青藏高原的绿色原野去。”雁说。

“你告诉我你能一口气从海上飞到世界最高山峰的另一边去？那一定很累吧！你飞过去大概要多久？”

“我估计飞过去要七个小时左右。返程的话能缩短三个小时！”

“I don’t know when you like to fly, but I take off at dawn. And then I do a straight run: from the sea to the green fields of the Tibetan Plateau,” Goose says.

“Are you telling me you fly from the sea to the other side of the highest mountain range in the world, in one go? That must be exhausting! How long does it take you?”

“I’d say about seven hours going up and over. The return is a quick three hours!”

“什么？你飞过去只花七个小时！这怎么可能！”

“为什么你认为自己做不到的事情就是不可能的呢？不如我们来为这件完全有可能做到的事情庆祝一下吧。”

“你的确称得上是‘鸟中宇航员’了！我很荣幸能成为你的朋友！”

……这仅仅是开始！……

“What? Going up and over takes you only seven hours! That is not possible!”

“Why do you consider something you cannot do, impossible? Let’s rather celebrate the fact that it is indeed possible.”

“You clearly deserve to be called ‘the astronaut bird’! I am proud to call you my friend!”

… AND IT HAS ONLY JUST BEGUN!…

……这仅仅是开始！……

... AND IT HAS ONLY JUST BEGUN! ...

你知道吗？

Common cranes breed in swamps, bogs and wetlands. They require quiet, safe environments, with minimal human interference. They occur at low density, ranging from only one to five pairs per 100 km^2.

一般来说，鹤在沼泽、泥塘和湿地中繁衍。它们需要没有人类干扰的，安静且又安全的环境。鹤的集群程度很低，每一百平方千米往往只能发现1–5对。

In winter, cranes eat the waste grain from harvested fields. Farmers benefit by having their fields cleared for use the following year, as well as by the cranes fertilising the soil with their droppings. These contain nutrients for microorganisms that are enriching the soil.

在冬季，鹤会吃收获过的农田里残余的粮食，它们清理农田便于农民来年耕种，它们的排泄物也会为土地增肥，农民们因此受益。鹤的排泄物中含有微生物需要的营养，因此能让土壤更加肥沃。

鹤在保护自己的时候，会用喙部啄，用翅膀拍打或者用爪子抓对方。为了躲避攻击，它们会飞到空中。当面对野猪或者狐狸这样的捕食者的时候，鹤甚至会先发制人地发起攻击。

When defending themselves, cranes jab with their bills, hit with their wings and kick with their claws. They avoid attacks by jumping into the air. They even deploy pre-emptive strikes against predators like boars and foxes.

大多数鹤分布在斯堪的纳维亚和俄罗斯。多数野生的鹤濒临灭绝，但是近年来这些鸟类的种群在德国、爱尔兰、英国、波兰和捷克共和国都有所恢复。杀虫剂中毒一度是鹤曾经面临的最大威胁，现在终于消失了。

The vast majority of cranes occur in Scandinavia and Russia. The common crane came close to extinction in the wild, but recently birds have repopulated Germany, Ireland, UK, Poland and the Czech Republic. Pesticide poisoning, once the greatest threat, is finally diminishing.

While many birds migrate over the Himalayas by using the valleys, bar-headed geese migrate from sea level in lowland India across the Himalayas to breed on the Tibetan Plateau. They fly non-stop to cross the mountains in as little as seven hours.

虽然许多鸟类利用谷地飞越喜马拉雅山脉，斑头雁会从印度低地的海平面飞越喜马拉雅，到青藏高原上繁衍生息。它们仅用七个小时就能一口气飞越整座山脉。

Bar-headed geese have a strict hierarchy, with mothers trying to secure a better future for their offspring. Lower-ranking females will attempt to lay their eggs in the nests of higher-ranking females, thus securing a jump up the social ranking.

斑头雁有着严格的等级制度，斑头雁母亲们会努力为自己的孩子营造更好的未来。低等级的雌性斑头雁会尝试在高等级斑头雁的巢里产卵，以确保后代的社会地位有所提升。

Bar-headed geese went locally extinct in many parts of the world. The species is, however, now gradually recovering. Scientists are very interested in their haemoglobin, which absorbs more oxygen into their blood than that of any other geese.

斑头雁在全世界很多地区都出现过局部灭绝，但是现在其种群正在逐步恢复。科学家们对斑头雁的血红蛋白很感兴趣，其吸收氧气进入血液的能力远超其他雁类。

The Ruppell's griffon vulture (Gyps rueppellii) is the highest-flying bird in the world. This species lives in the Sahel Region of North Africa, and is, unfortunately, now critically endangered. Due to loss of habitat, only 30,000 are left.

鲁氏粗毛秃鹫（黑白兀鹫）是世界上飞得最高的鸟类。该物种生活在非洲北部的撒赫勒地区，不幸的是，它们现在极度濒危。因为栖息地消失的原因，全世界只剩下30 000只了。

Think about It

When you are told, "It is impossible", do you accept that it is impossible?

当有人告诉你“这不可能”的时候，你会认为这就是不可能的吗？

Is it that easy to fly over the Himalaya Mountains by breathing deeper?

加深呼吸会更容易飞越喜马拉雅山吗？

Is there always someone who will be better than the best?

总是会有一个比“最优秀的人”还要优秀的人吗？

What about flying across the Himalayas only eating grass?

你怎么看只靠吃草就能飞过喜马拉雅山这件事？

Are you aware of your strengths ? List the things you are good at, which are usually also what you enjoy doing most. This knowledge will help you in achieving success in your career. If you are, for instance, good at playing a musical instrument, then precision is one of your strengths, as may be your ability to focus, and to listen carefully. That is three useful skills right there! Share your list with friends and family members, and ask them if they agree. They may even add some items to your list. Now the most important step is to find out which professions require these skills. Have fun charting your future!

你了解自己的强项吗?列出你擅长做的事,这些往往也是你最喜欢做的事,这有助于你在事业上获得成功。例如,假设你善于演奏一种乐器,那么细致就是你的长处之一,也许你还拥有专注力并且善于倾听。这就算三项有用的技能了!把这个清单分享给你的家人和朋友,问问他们是否同意。他们甚至可能会再补充几条。接下来最重要的是找到需要这些技能的都有哪些行业。祝规划未来愉快!

学科知识

Academic Knowledge

生物学	捕食者在生态系统的功能至关重要；鹤的左心室具有更多毛细血管来加速氧气供应；鹰飞行时高度最高可达海拔5 000米，速度可达每小时150千米；几乎所有主要的动物群体都会迁徙，包括鸟类、哺乳类、鱼类、爬行类、两栖类、昆虫类和甲壳类。
化　学	每立方米的分子数量是固定的，所以当有更多水蒸气的时候，每立方米气体的分子质量就会下降；血红蛋白对氧气的亲和性很高，但在高浓度二氧化碳下会释放氧气。
物　理	密度是物体的质量除以体积，并且由温度、压强和水蒸气含量决定；温度越高，分子运动越快；空气密度随温度升高而降低；气压降低会导致温度降低；气压降低减少氧气含量；潮湿的空气比干燥的空气轻（密度小）；地速与空速不一样。
工程学	赛车的设计像倒置的飞机机翼，可增加抓地力，空气密度低的时候，由于升力小，飞行器需要更长的起飞滑跑距离；涡轮增压器能增加进入发动机的空气密度，使得车辆有更大功率，飞行器有更强的升力。
经济学	管理气压和温度可提高能源利用率，节约成本；反弹效应：越来越多机构承诺清除垃圾，但因为越来越多的人不处理垃圾，总量仍在上升；有人通过收集垃圾换取金钱。
伦理学	对捕食者的伤害威胁到了海洋和陆地的生态环境，影响食物产出、疾病控制和气候稳定。
历　史	牛顿在1717年指出，潮湿的空气比干燥的空气密度低；阿伏伽德罗发现用水蒸气取代氮和氧会降低空气密度；马奎斯·德·阿里安德斯在1783年用热气球实现了人类的首次飞行；珠穆朗玛峰。
地　理	北极燕鸥为享受两个夏季每年迁徙71 000千米；斑头雁从印度北部飞到南部过冬。
数　学	空速、地速和风速都是矢量。
生活方式	呼吸模式对大脑影响深远，缺乏锻炼对大脑和身体健康都有影响；深呼吸不会给大脑充氧。
社会学	一些地区或民族的人认为把死者留在珠穆朗玛峰上是对山神的不敬；还有一些地区或民族的人认为珠穆朗玛峰是神圣的。
心理学	抵抗干扰需要意志力；呼吸小窍门：小幅度吸气后立即放松可以延长呼气；有志者，事竟成。
系统论	捕食行为比自上而下的食物循环更重要；植物生长茂盛是因为食肉动物控制食草动物的数量。

情感智慧

Emotional Intelligence

鹤

鹤邀请朋友一同旅行，他真诚地介绍了自己的习性。对于到达繁殖地，鹤有很强的动力。他在解释人们做不到而他却做得到的事情时提供了相关知识。他清楚自己的极限，并不嫉妒秃鹫比他飞得更高。他很难相信雁的飞行能力，甚至声称那是不可能的。当雁为此批评鹤，并让鹤为她的确有强大飞行能力而庆祝的时候，鹤通过赞赏雁的飞行能力，快速地加深了他们的友谊。

雁

雁了解她的朋友，并分享了她的见解。为了到达繁殖地，她有着巨大的动力，并许下了强有力的承诺。她清楚认识到需要面对的挑战，尤其是氧气需求，但是她准备好尽力而为。她在对话中增加了有关人类行为的信息。她敬佩秃鹫，但也为自己和鹤能做到的事情而感到自豪。她乐意分享自己的饮食习惯和迁徙方式。对于鹤认为雁无法如此快速地飞行这件事，她感到沮丧，希望鹤为雁的确能做到这件事而庆贺。

艺术

The Arts

唐卡是传统藏式艺术。让我们发挥创造力画出不同图案吧！你需要一块方形的纯棉织布，以及一些水溶性的布彩颜料。唐卡的风格非常几何化，比如说画人的时候，四肢和五官都需要按照系统性的网格角度和相交线来排布。正规的唐卡绘画有相当多的规则。你不需要遵守这些，但是遵循这些方法的话你会学到很多关于设计和对称的知识。如果你精通了这种绘画风格的话，你可以轻而易举地在周围的旧T恤衫上作画。

思维拓展

Systems: Making the Connections

最初没人能预料到成堆的垃圾和尸体会被遗留在高山上。更糟的是高山环境难于清理，低温又保存了这些遗弃物。如今人们认识到了对环境造成的诸多负面影响，例如人兽冲突、捕食者数量锐减引起的生态失衡。生态系统需要顶级捕食者保证物种的动态平衡。这就像是一个自我调节系统，删除其中任何一环（比如海胆-海带关系中的海獭，或者麋鹿-河湾物种生长关系中的狼）都会导致其余环节的改变。

科学家过于关注单一物种，却忽略了食物链顶端剧烈变化带来的紧密影响。猎物知道如何规避遭遇捕食者的危险，人们需要更好地理解捕食者和猎物的关系。丢弃垃圾、引入新物种、破坏生态链平衡的行为就像蝴蝶效应，需要投入更多精力去补救。

这就是研究雁与鹤的飞行给我们带来的思考。它们在效率和速度上都优于其他鸟类，而我们最多只能做到标记这些鸟类，并追踪它们的路线。现代社会需要把关注的重点从保护栖息地转为积极恢复生态系统，即关注所有生物，从土壤中的生命到食肉动物，在生态系统中的作用。

动手能力

Capacity to Implement

大多数纸飞机都被设计为可以在空中滑翔。为了说明对抓地力的理解，我们来做几个纸飞机：雁快速离地飞行和猎豹高速转弯而不减速时用的力是一样的。为了展示雁对抓地力的应用，你的第一架纸飞机在发射时应采用陡峭的上抛轨迹。只要我们遵循了基本原则，纸飞机是否立即下跌并不重要。接下来，设计一架纸飞机来展示猎豹对抓地力的应用，发射时需要俯冲。研究这两种方法背后的几何学知识。你已经迈出了成为流体动力学工程师的第一步！你甚至某一天可以加入那些把卫星送入轨道的团队，或者是把加速时的（电动）赛车送入弯道的团队。

故事灵感来自

This Fable Is Inspired by

朱莉娅·M·约克

Julia M. York

朱莉娅·M·约克出生于加拿大不列颠哥伦比亚省。2015 年她毕业于不列颠哥伦比亚大学，获生物学学士学位。2012 年至 2013 年，她在乌普萨拉大学学习了一年的分子生物学。2016 年，她继续攻读硕士学位。朱莉娅决定在奥斯汀的得克萨斯大学继续科研工作。她的目标是在 2022 年获得生态学、进化和行为学博士学位。朱莉娅自认为是一名比较生理学家，她对用神经科学来回答进化问题很感兴趣。她正在研究热感觉、生理极限和气候变化影响之间的关系。当她还是不列颠哥伦比亚大学的一名本科生时，为了研究斑头雁的呼吸生理，她带着口罩和心率监测器训练它们在风洞中飞行。

图书在版编目(CIP)数据

冈特生态童书.第七辑:全36册:汉英对照 /
(比)冈特·鲍利著;(哥伦)凯瑟琳娜·巴赫绘;
何家振等译.—上海:上海远东出版社,2020
ISBN 978-7-5476-1671-0

Ⅰ.①冈… Ⅱ.①冈… ②凯… ③何… Ⅲ.①生态
环境-环境保护-儿童读物—汉英 Ⅳ.①X171.1-49

中国版本图书馆CIP数据核字(2020)第236911号

策　　划　张　蓉
责任编辑　程云琦
助理编辑　刘思敏
封面设计　魏　来李　廉

冈特生态童书
鸟中宇航员
[比]冈特·鲍利　著
[哥伦]凯瑟琳娜·巴赫　绘
贾龙慧子　译